GEM

A Guide to Colorado's Native Gemstones

Lee & Tag McKinney

BY LEE & TAG McKINNEY

ISBN: 1-55838-072-8

RENAISSANCE HOUSE
A Division of Jende-Hagan, Inc.
541 Oak Street ~ P.O. Box 177
Frederick, CO 80530

Cover and all Interior Illustrations
by the Authors

10 9 8 7 6 5 4

Contents

NOTE: References to sizes of specimens below photographs are in centimeters.

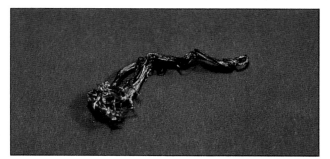

Gemology is the science of minerals and other material beautiful and durable enough to use for personal adornment. Of the 2,400 or so mineral species, about 100 fit the above critera to be called gemstones. Colorado has more than 400 identified minerals.

A rock is a geologic unit made up of one or more minerals, a mineral being a natural chemical substance, with a definite crystalline structure. Rocks and minerals are formed by a fusion process (*igneous*); deposited from a solution (*sedimentary*); or alteration of existing rock (*metamorphic*).

There are three divisions of minerals: GROUP (garnet and feldspar); SPECIES (quartz, beryl, etc.) with the same chemical compostition, hardness, specific gravity and refractive index; VARIETY, a subdivision of species differing in color, form, transparency or phenomenon. Emerald and aquamarine are varieties of the species beryl; sapphire and ruby of the species corundum. What distinguishes gems from minerals is that the latter are made by nature, unaided by man. Gems are usually minerals changed by carving, cutting into cabochons (smooth domes), or faceting (cutting at angles).

Aquamarine, the state's gemstone, found on Mt. Antero, probably has the greatest monetary value of any Colorado mineral except gold. Amazonite, the state mineral, is found in the Pikes Peak region and valued throughout the world. Smoky quartz and amazonite combinations (picture p. 20) are uniquely Colorado and rarely found in other places.

The principal factors affecting value are beauty, durability and rarity. Rock shop or mineral dealers can usually suggest local collecting areas, clubs, mineral exhibits and museums. They also have examples of local material available.

Crystals: There are six *crystal systems:** Isometric, Tetragonal, Orthorhombic, Monoclinic, Triclinic, Hexagonal (including Rhombohedral). Each has numerous *habits* (external crystalline shapes).

Hardness: The relative *hardness* (H.) of minerals is measured on the Mohs scale ranging from talc which is one (soft) to diamond at ten (hard). Standard tests for hardness involve scratching one mineral with another. For example, calcite and a copper penny have the same hardness, 3, while a knife blade is harder than either one with a hardness of 5.5 and quartz harder yet with a Mohs scale rating of 7.

Following is a list of some of the well-known minerals and the counties in which they are found.

```
AGATE - Eastern & Western regions
ALABASTER - Larimer County
AMAZONITE - Front Range area
AMETHYST - Larimer and Saguache
ANATASE - Boulder, Gunnison
BARITE - Weld, Park, Mesa
BERYL - Chaffee, Fremont, Larimer
BIOTITE - El Paso, Fremont, Larimer
CALCITE - Mesa, Weld
CHALCEDONY - Gunnison, Mineral, Saguache
EPIDOTE - Chaffee
GALENA - Lake, Ouray, Park
GARNET (Almandine) - Chaffee
       (Grossular) - Chaffee, Gunnison
       (Spessartite) - Chaffee
GOETHITE - Park, Teller
GOLD - Boulder, Chaffee, Clear Creek,
       Gilpin, Lake, Park, Summit
JASPER - Chaffee, Park
JET - El Paso
LAPIS LAZULI - Gunnison
MOLYBDENITE - Chaffee, Lake, Summit
MUSCOVITE - Fremont, Larimer
OBSIDIAN - Chaffee
OPAL - Park, Mineral
PYRITE - Most mining areas
QUARTZ - Most mining areas
RHODOCHROSITE - Clear Creek, Gilpin, Park,
                Ouray, San Juan, Summit
RHODONITE - San Juan
SATIN SPAR - Fremont, Larimer
SILVER - Boulder, Chaffee, Clear Creek, Gilpin,
         Hinsdale, Lake, Mineral, San Juan, Summit
SPHALERITE - Ouray, San Juan
TOPAZ - Chaffee, El Paso, Park, Teller
TOURMALINE - Fremont
TURQUOISE - Conejos, Lake, Saguache
WOLFRAMITE - Boulder, Gunnison, Park, San Juan
ZIRCON - El Paso
```

*See *Textbook of Mineralogy* by E. S. Dana and W. E. Ford, for drawings of all six crystal systems and various habits in each.

AMAZONITE (Triclinic - 6 H.) - COLORADO STATE MINERAL

MUSEUMS and HISTORIC SITES

When Edwin Carter came to Breckenridge in 1870, he began collecting natural history specimens. By 1900 his collection was extensive enough to interest the newly formed Colorado Museum of Natural History in purchasing it as the nucleus of its collection. The east building extension and first viewing was in 1903, with the main museum building finished by 1908. Collections were acquired primarily by donation and one of the first was John F. Campion's crystallized gold in 1900. The Porter-Pratt minerals and cut gems were donated in 1912 and Pohndorf's in 1924.

Today the Denver Museum of Natural History houses one of the finest gem and mineral collections in the world. There is Campion's crystallized gold and "Tom's Baby" from Breckenridge (the single largest piece of gold ever found in Colorado); Rhodochrosite, a beautiful pink-red cubic mineral, from the Sweethome Mine at Alma in Park County. A pocket of Amazonite from the Crystal Peak area has been recreated to demonstrate a "pocket" in place as it is in nature. Many other high quality Colorado minerals and beautiful world-wide specimens are now housed at the Museum, which is open 9-5 daily. Spending some time in museums before going field collecting can educate the traveler as to the *size, shape,* and *color* of top quality gems or minerals from a particular locale. Ideally, the mineral will be on matrix to aid in

5

identification of the host rock. Both the Denver Museum of Natural History and the Colorado School of Mines Museum in Golden have good examples of AMAZONITE (the Colorado *state mineral*) and AQUAMARINE (the Colorado *state gemstone*). These museums contain two of the best Colorado gem and mineral collections in the state.

Colorado School of Mines, perhaps the world's foremost college of mineral engineering, has a replica of a mine, various mining machinery, lamps and other memorabilia from the mining industry, as well as a wonderful collection of historic scales and items used in an assay office. The world-wide mineral collection in their museum emphasizes Colorado material. There are also cut stones from various locations, in beautiful displays.

The Colorado History Museum at 1300 Broadway in Denver has a fine exhibit on early mining, as does the Western museum of Mining and Industry near the Air Force Academy, north of Colorado Springs. There are mining museums in Victor, Telluride and other old mining towns. Leadville houses the Mining Museum and Hall of Fame.

Just north of the Coors plant in Golden is North Table Mesa mountain, a location famous for zeolites--groups of hydrated silicates of aluminum with alkali metals. This is a private area but permission is sometimes granted for experienced collectors.

At the Morrison exit off I-70 is the historical marker noting 20 million years of earth's history, from the Jurassic Period 140 million years ago, to the Cretaceous, 120 million years ago. Seventy-five million years ago, in the late Cretaceous period, a mountain range was born, pushing up this Hogback ridge. A path has been cut above the highway, allowing you to walk through this incredible period of history and study the geology from informative plaques. In western Colorado, the visitor center at Colorado National Monument offers interpretive programs and exhibits to explain the fascinating geologic history of Western Colorado.

BLUE BARITE (Orthorhombic 3-3.5 H.)

STONEHAM

You can find Stoneham barite, perhaps the finest blue barite in the world, on the Nicklus Ranch, just east of Stoneham. A blue color on the surface of the ground, or partial crystals (flecks) of barite are good indications of where to begin your hunt; previous diggings in the area are also a helpful guide. Find a calcite seam (white to cream color) and follow it or look for the crystals on the surface of dry stream beds. Most of the crystals are singles (1/4" x 1"), but large plates of barite have been found here, with crystals 1" x 4" long.

This barite is tabular in form and is translucent to transparent, a medium blue color. It cleaves (separates) easily and has a high specific gravity (is heavy). Barite crystals are found both on the surface and many feet underground. They are highly prized as mineral specimens world-wide.

Stoneham is west of Sterling 25 miles on Colo. 14. East of the jct. of Colo. 14 and 71, take the first road north. About two miles on left is the Frank Nicklus Ranch. Ask permission to dig, then travel north to the oil storage tanks, turn right on road leading to clay bluffs. Park on south rim and walk down into gulley.

MINING MEMORABILIA

RED FEATHER LAKES

Site of AMETHYST (Rhombohedral - 7 H.)

Red Feather Lakes, northwest of Fort Collins, is an area of gem quality amethyst, a beautiful purple-violet variety of quartz. This is a private fee area, where the traveler must pay to gem hunt on private land. Look along the creek bed for "float"--a mineral not in place due to erosion or geologic upheaval--which is usually downhill from the source. Quartz crystals are typical six-sided crystals with pointed tops. Amethyst mineral specimens from this area are sharp crystals with good terminations (pointed tops) and excellent color. While in the field, don't overlook partial or broken crystals for friends who facet or do cabochons.

In the breastplate of Aaron, 1st High Priest of Hebrews, amethyst is one of the gemstones representing the 12 tribes. Centuries later it was assigned to one of the 12 apostles and later still as the birthstone for February, one of the 12 months. Although plentiful, it is one of the most beautiful gemstones in the world.

> Take U.S. 287 north from Fort Collins to the Forks, turn left onto Colo. 200 west to Red Feather Lakes. Stay on pavement until it ends, then continue four miles, across a creek. Keep right to the Rainbow Amethyst Mine.

CRYSTAL MT. (Pocket in place)

CRYSTAL MOUNTAIN
Site of BERYL (Hexagonal - 7.5 H.)

West of Fort Collins 45 miles is a large six square-mile area of pegmatites known as the Crystal Mountain area. A *pegmatite** is a very coarse-grained granite formation which crystallizes from solutions flowing out of nearly solidified granitic masses. Sometimes these solutions contain exotic elements *not* crystallized in the granite. These elements form the rare minerals in this area such as: autunite, bertrandite, bismuthinite, bismutite, columbite-tantalite, fluorapatite, muscovite, spodumene, triplite, torbernite and uraninite. These minerals are usually within miarolitic cavities (gas bubbles) called "pockets" (see p. 15) associated with quartz and beryl. The pegmatites of Crystal Mountain, once mined for beryllium ore, are mineralized with opaque, sharply crystallized beryl, making good reference collection material. Look in the granite boulders and outcrops for small pockets containing crystals--quartz, beryl or any of the above-mentioned minerals. The species beryl includes the following varieties: aquamarine (blue), emerald (green), goshenite (colorless), heliodor (gold), morganite (pink), and red beryl.

Take Horsetooth Road west from I-25 to Masonville. Continue west on Buckhorn Creek Road approximately 18 miles to Crystal Mt. turnoff. Go south 2+ miles to Crystal Mt.

*For more on pegmatites, see p. 21

SILVER 3x4   CRYSTALLINE GOLD at bottom (both Isometric 2.5-3 H.)

BOULDER/WARD/NEDERLAND

The 40th Parallel (now Baseline Road in Boulder) once divided Nebraska and Kansas Territories, both of which recorded mining claims. Gold in the hills behind Boulder was discovered in 1859, on Gold Run between Four Mile Creek and Left Hand Canyon. Soon, mining towns like Gold Hill, Summerville, Salina, Wall Street, Sunset, Sugarloaf, Magnolia, Sunshine, Caribou, Jamestown, Ward, Nederland, and Eldora sprouted up throughout the area. Some of these towns still exist, while others are merely remnants of the mining heyday. Several short excursions take the traveler or collector into the Boulder Mining District. These trips provide an excellent tour of gold mining country, and are a delight for rockhounds, collectors, or the general traveler looking simply for a scenic, historic drive.

One-hour trip: From Broadway and Canyon in Boulder, travel Colo. 119 up Boulder Canyon 2.7 miles, turning right onto Boulder Co. Rd. 118. Notice the mill foundation at 7.1 miles. Turn right onto Co. Rd. 89 (sign to Gold Hill) at 7.5 miles. The Old Salina church is at 8.5 miles and mine tailings start thereafter. One mile further is Summerville, with dumps above the town and a great view of the plains below. At the top of

10

the ridge is a shaft, with dumps spotting the ridge. Gold Hill at 11.6 miles is home to The Blue Bird Lodge, a bed and breakfast, where Eugene Field wrote *Casey's Table d'Hote*. The town was first destroyed by fire, horror of all early Colorado towns, in 1860 and next in 1894. A road from here to Ward takes the traveler to the Peak to Peak highway (Colo. 72) and back to Boulder via Nederland and Sunshine, down Sunshine Canyon.

For an alternative route back to Boulder, turn right out of Gold Hill to Co. Rd. 52, viewing mine dumps all the way to the top of the ridge and down the other side. About four miles from Gold Hill, the road intersects with pavement which emerges in Boulder on Mapleton Hill, site of beautiful historic homes built during the mining boom of the late 1800s. The University of Colorado Henderson Museum's Geology Exhibit makes an educational interlude between tours.

One and one-half hour trip: Head north on U.S. 36 to Co. Rd. 94 (Lefthand Canyon) and on to Jamestown about 15 miles, noting mine tailings just above the town. Proceed along the Peak to Peak highway (Colo. 72) to Ward, a boom town of 600 by 1865, which produced $5,000,000 in gold from the Columbia claim alone. Both then and now, nearly any area stream can be panned and "color" will appear. The narrow gauge railway from Boulder to Ward was finished in 1898, and an average of 250 passengers and 100 tons of ore were transported per day. Fire destroyed most of Ward in 1900.

Nederland, 14 miles from Ward, had a mill which concentrated silver ore from the Caribou and other mines to make silver bricks. From Nederland, the traveler can return to Boulder (20 miles) or go on to Central City (15 miles).

When General Grant visited Central City in 1873, the pavement in front of the Teller House was made of these bricks because gold was so common it was not considered worthy. Silver, which is more abundant than gold, sometimes forms spectacular crystals, wires and sheets. Silver occurs in veins with gold, or with ores of zinc and lead. But the big discovery in 1900 was "black iron" (tungsten), which is wolframite and huebnerite (Monoclinic, 4-4.5 H. red-brown to black--see rhodochrosite picture p. 27).

CHALCOPYRITE 4x7   Glory Hole (Tetragonal 3.5-4 H.)

GILPIN/CLEAR CREEK AREA

Head west from Golden on U.S. 6 through Clear Creek Canyon to Colo. 119, to Blackhawk and Central City.

CENTRAL CITY/BLACKHAWK--These two old mining towns are steeped in mineral history from 1859. Blackhawk had most of the smelters and mills; Central City, most of the entertainment. There are performances each summer at the Opera House, and at the nearby Teller House is the famous "Face on the Barroom Floor." During the summer months, fee areas along Clear Creek are open for gold panning. Gold, the major mineral in the area, is still being produced on private claims. Rock shops will have specimens.

Take gravel road 281 from Central City via Virginia Canyon, former home of hundreds of mines, to Idaho Springs.

IDAHO SPRINGS--The Underhill Museum on Miner Street is a good place to begin. There are guided walking tours of the silver-producing Lebanon Mine. At the Argo Mill, the traveler can see how gold bearing ore (p. 18) was processed, and view the tunnel which was used for water drainage and transporting ore to the mill. The Colorado School of Mines offers tours of its experimental Edgar Mine during the summer months. West from Idaho Springs on the south side of Clear Creek is the old Stanley Mine, operating now for gold and silver.

PYRITE 9x12 - Waldorf (Isometric 6-6.5 H.)

The Fall River Road (275) two miles west of Idaho Springs toward St. Mary's Glacier leads about 8.5 miles to the left turn into Alice. Cross Silver Creek Road, proceed up hill to the Glory Hole.

ALICE--A yellowish host rock can be seen on the left going into Alice. An open pit mine is on the right, but use caution--overhangs are very dangerous. Nice little plates (groups) of quartz, pyrite and chalcopyrite have been found in this area. Pyrite, commonly known as "fools gold," is much harder than gold. Crystals are very common throughout the world. Chalcopyrite resembles gold and pyrite in its brassy yellow color but is harder than gold and softer than pyrite. The mineral frequently tarnishes to a beautiful iridescence. Both chalcopyrite and pyrite are most attractive when associated with white to clear quartz crystals.

From Idaho Springs, continue west on I-70 to the Georgetown exit.

GEORGETOWN--The silver craze was well underway in 1867 when Georgetown (Silver Queen of the Rockies) produced the Anglo-Saxon lode whose ore assayed at $23,000 a ton. Today, any mine dump in the area has potential for a nice silver specimen or at least an ore sample. Georgetown is now a haven of restored Victorian architecture.

Road 118 up Mt. McClellan south of Georgetown ascends switchbacks nearly to Clear Lake. At the fork, turn right and proceed five miles to the old site of Waldorf.

WALDORF--Excellent pyrite and other minerals can be found on the dumps surrounding Waldorf. The Santiago Mine at 12,000 feet usually produces small but nice mineral specimens of pyrite groups sometimes associated with quartz druse (tiny terminated quartz crystals). A scenic return to Denver is over Guanella Pass to Grant and U.S. 285 to Denver.

13

SMOKY QUARTZ 6.5x15   Devilshead (Rhombohedron - 7 H.)

DEVILSHEAD

Go south from Denver on U.S. 85 to Sedalia. Turn west on Colo. 67 toward Deckers. At the ranger station, head south on Rampart Range Road 11.2 miles to Devilshead and south to the picnic area called Topaz Point. Back up 300 feet to a road on the left, which leads to the Colorado Mineral Society claims of smoky quartz and topaz. Look on either side of this road.

Old timers know this area as Virgins Bath because thousands of years ago, a natural "bathtub" of granite was formed here. Traces of a red iron-stained mud coming from the pegmatites and/or a small seam or crack may lead to a "pocket" of smoky quartz. Pockets containing over 100 crystals and longer than 12 inches come from this area, but a two-inch specimen is respectable. Smoky quartz cuts beautiful gems so take the partial crystals home for your friends who like to facet.

Most of the Devilshead topaz has been found on the right side of the road, but it is possible to find throughout the area. Topaz crystals as large as a baseball are known here. This smoky quartz and topaz are fine mineral specimens and make excellent, clear faceting material. Pink feldspar (p. 17) and amazonite (p. 20) are also found here.

14

TOPAZ 2x3   blue - Devilshead (Orthorhombic - 8 H.)

LONG HOLLOW

South from Devilshead on the Rampart Range road 7+ miles is the entrance to Long Hollow, where some of the largest smoky quartz and topaz crystals in Colorado were found. The smoky weighed 44 pounds and the topaz more than 17. Down the hollow on the south side, large topaz crystals are embedded in the granite faces. The finest topaz in this area, clear or pale blue, is in cavities that have iron-bearing mud and sand at the bottom. These cavities range from baseball-size to a foot or more in diameter.

Crystals are usually found in a gas bubble area called a "pocket" (miarolitic cavity) within a pegmatite dike (p. 8). These pockets are sometimes collapsed, meaning that their walls, with crystals attached, have fallen during geologic upheaval, causing some crystals to be "dinged" (chipped or broken) or causing cleavages (separations always parallel to a crystal face). Pockets range from one inch to a cavity in which a man can stand upright. When digging in these mud-filled collapsed pockets, beware of quartz slivers in the mud which can cut fingers. The patient and persistent may be rewarded with perfect little plates.

ZIRCON crystal 1x1.5   St. Peters Dome (Tetragonal 6.5-7 H.)

ST. PETERS DOME

From Colorado Springs, take Gold Camp road past Seven Falls through tunnels #1 and #2 where the road widens. Informal parking area on right. Begin looking on the downhill side of the road.

St. Peters Dome is a fine area to search for sharp opaque brown zircons. Most of the zircons will be in a quartz matrix and will require a dental pick to remove the material around them. These zircons are not facet quality, but they make excellent, sharp mineral specimens, looking like squares balanced on a corner point. Most zircon crystals are less than one-quarter inch but three-quarter-inch specimens come from this area. To reach the Old Eureka tunnel locality for these zircons, continue to a jeep road on the left. Stay right down the hill 1/4 mile to the old tunnel. Red clay streaks appear in some old diggings above the road near the top of the ridge.

In this otherwise barren area, there may be nice specimens in Pikes Peak Granite, a graphic granite of feldspar-quartz intergrowth sometimes with mica and usually decomposed.

Some minerals common to the region are listed below. These specimens range from 1/4" to 6" for a single crystal.

Amazonite, blue-green, slightly more blue than turquoise (p. 20)
Goethite, black (p. 21)
Smoky quartz, clear brown-black (p. 14)
Pink feldspar, salmon pink (p. 17)
Fluorite, usually lavender but sometimes pale green (p. 32)

PINK FELDSPAR 12x15   Harris Park

CRYSTAL PARK

Take Sutherland Creek Road SW from Manitou Springs to Cameron Cone. Consult Forest Service map for this area. Much of Crystal Park is now a private area.

Crystal Park, one of the oldest known collecting localities in Colorado, is based at approximately 8,500 feet in Pike National Forest. This area has amazonite, clear and smoky quartz, phenakite and topaz. These specimens will occur in cavities of the pegmatites in Pikes Peak granite. Phenakite crystals, mostly colorless, may at first glance look like doubly terminated clear quartz. The famous white stripe amazonite found by Clarence Coil, noted early Colorado mineral collector, comes from this area. Topaz crystals as big as a fist have been found here, not always in pockets but sometimes in float (p. 8). These minerals may be found in any outcrop along the Front Range which contains mineralization. Killer specimens, collected for their beauty, are possible at any point.

It takes practice to identify a specific mineral, and even more skill to find a particular one. Just being able to recognize a specific mineral without reading the label in a rock shop is quite an accomplishment.

GOLD in QUARTZ 4.5x7   Russel Gulch

CRIPPLE CREEK/VICTOR

West from Colorado Springs, take U.S. 24 to the town of Divide. Turn south on Colo. 67 toward Cripple Creek and Victor, driving along the western edge of Pikes Peak.

The Cripple Creek/Victor area is one of the richest gold producing areas on earth. During its peak years, Cripple Creek's population was 35,000. In April, 1896, a fire destroyed much of the oldest part of town, including town records and newspaper archives. In one year, 1900, output from the mines in this area has been variously recorded from $18 to $23-million. The Mary McKinney Mine alone produced $12-million in a single year. Just south of Cripple Creek is the town of Victor, which at its peak boasted a population of 17,000.

In the early days, most mining was from placers. It was not until 1893, after millionaire Winfield Scott Stratton had made his strikes at the Independence and Washington Mines, that the big underground mines of the district were developed. Today there are many active mines in the Cripple Creek/Victor area. All mines are on private property, so field collecting is impossible without a contact. Extreme caution is advised in exploring the area, as open mine shafts can be very hazardous. Leaching fields, seen today between

Cripple Creek and Victor, illustrate how mining methods have changed through the years. Today, chemicals are used to leach gold and other minerals from crushed rock, rather than the gold panning methods of earlier years.

Gold, which caused all the furor in this area, has been valued by civilizations for centuries, and is used today as a monetary standard in many countries, and in crafting jewelry. Pure gold is 24 karat, a designation of fineness. Specimens less than 24 karat are not pure gold, but have been mixed or alloyed with other metals. The color of the gold depends on the amounts of other minerals associated with it. Copper, for example, will make gold appear more pink, nickel more white. Silver is often found in the same areas as gold. Pure silver is 99.99% silver (or 999 fine). Sterling is not less than 92.5% silver (925 fine), while coin silver is generally 900 fine.

Summer travelers to the Cripple Creek/Victor area can tour a mine or ride the Cripple Creek & Victor Narrow Gauge Railroad. Victor maintains an exquisite museum and associated assay office, where much of the apparatus for assaying is on display. Many antique shops in the area sell memorabilia such as carbide lamps, mining stocks, ingot molds, muffle furnaces, etc. The Cripple Creek District Museum offers a great education on both the social and mining history of what has been called "The World's Greatest Gold Camp."

AMAZONITE & SMOKY QUARTZ 9x14   Harris Park

FLORISSANT AREA

West from the town of Divide is Florissant and the Fossil Beds National Monument, a spectacular area for rockhounds and travelers interested in geologic history and education. There is no collecting at a national monument. North from Florissant is Crystal Peak, famous worldwide for Colorado amazonite, the state mineral. There are many private mining claims in this area requiring permission to collect, as well as open areas with outcrops of pegmatites where the sharp-eyed rockhound may unearth a beautiful plate of amazonite and smoky quartz crystals, the classic combination. Amazonite is the blue-green variety of microcline in the feldspar group. Feldspars are very common in the earth's crust but this well-crystallized blue-green variety is found only in a few places. Colorado--Crystal Peak in particular--is a classic location.

LAKE GEORGE AREA
Site of GOETHITE

Look for a seam in the Pikes Peak granite pegmatite. Usually red clay in the seam leads to a mineralized pocket. The goethite pictured is from a pocket containing 127 specimens ranging from an inch to large plates. Goethite is shiny black and often radiates

GOETHITE 6x10   Lake George (Orthorhombic 5-5.5 H.)

needles in a circular pattern. It is an iron ore mineral and may oxidize, giving a rusty appearance.

Smoky quartz and pink feldspar are often associated with goethite in this area, making outstanding multi-mineral plates which are prized by collectors. A beautiful pocket of goethite associated with amethyst quartz crystals was found recently, but this combination is most unusual.

Turn right at the Lake George ranger station, head north one mile to a small road on the left (the first one past the city dump) to the collecting area. Take it two blocks to an obvious parking area.

PEGMATITE*

Pegmatite refers to any exceptionally coarse-grained phase of rock. Pegmatite bodies in this area occur within granite. Small bodies enclosed in granite are believed to be places where accumulation of gasses prevented normal solidification of the rock, creating openings where fluids and gasses migrated. The term *miarolitic* describes pegmatitic granitic rock honeycombed by gas openings.

Granitic pegmatites comprise the principal parent magma where feldspar is predominant, with abundant quartz and mica. In the front range, minerals are usually found in miarolitic cavities ranging from one inch to several feet in the Pikes Peak red granite region.

*For more on pegmatites, see p. 8

TARRYALL AREA

Topaz crystals can be gigantic--over 100 pounds in some parts of the world--but not in Colorado. Our size is smaller and our colors pastel instead of the dark hues from Brazil, the most popular shade being yellow-brown. Because quartz is much less expensive than topaz, there have been many topaz imitations from the quartz family. "Smokey topaz" is really smoky quartz; "Bohemian topaz" is really citrine (yellowish quartz). Topaz is, in turn, used to imitate aquamarine. The pale blue is often irradiated to a darker blue resembling aquamarine and the clear has even been used for diamonds in years past! All of these stones are lovely when faceted into gemstones but: *caveat emptor!*

Topaz is heavy compared to quartz or aquamarine and could be distinguished by heft. The Colorado topaz has a high degree of brilliance and can be colorless, pale blue, pale yellow, or sherry brown, all gorgeous when cut but spectacular also as terminated mineral specimens, particularly when associated with other minerals.

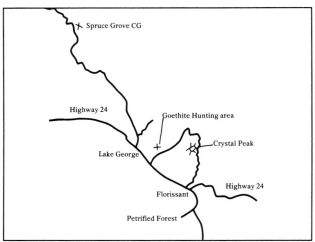

From Lake George, cross the South Platte River. Turn right toward Tarryall Reservoir. Park at Spruce Grove campground a few miles before the reservoir. Cross Tarryall Creek by footbridge and hike up the hill one-fourth mile. Take the right fork when the trail divides and follow the ridge about a half mile to the Colorado Springs Topaz Claim, sloping down the hill on the east side.

TOPAZ longest, XL 4x4.5   Tarryall (Orthorhombic 8 H.)

In hunting topaz, find a likely hole among the many existing or start your own. Topaz is sometimes found as float (p. 8) in the loose dirt above the pegmatite. More often, though, the collector should look for a small crack with red mud, and follow it until the pegmatite opens to a nice pocket.

The left fork at the top of the hill goes to the Moore Claims, open to field collectors. Follow the paint marks on the trees through the forest to the steep uprise of the Tarryall Mountains with jagged rocks to the northeast.

Digging a foot or less into the gentle slopes of the Moore claims in front of the Tarryall pinnacles can produce float (topaz). For technical rock climbers, the pinnacles will yield topaz in small vugs (holes often lined with crystals), probably the primary source of this topaz. Topaz cleaves easily so the collector should look for parallel cleavages or one crystal face which will slick clean with thumb pressure up the side. Quartz breaks with a conchoidal (shell like) fracture but topaz breaks even and smooth. Beautiful terminated water-clear crystals of 500 or more carats--as large as a coffee cup-- have been found in this area. Crystals range from this large variety down to the size of a pea, in both blue and clear shades.

23

Pg 10 - Boulder/War
GOLD

Pg 12 - Central City/Blackl
GOLD, SILVER

Pg 13 - Georgetown/Waldorf
QUARTZ, PYRITE, SILVER

MARYLAND CartoGraphics, Inc.
Columbia, Maryland 21045

COLORADO MINERAL LOCALITIES

NOTE:
BOLD ITALIC indicates mineral bearing rock or mineral artifacts

Pg 27 - Alma (Sweethome Mine)
PYRITE, FLOURITE, RHODOCHROSITE

Pg 28 - Leadville
BARITE, DOLOMITE, QUARTZ, GOLD, PYRITE, TURQUOISE, RHODOCHROSITE, SILVER,

Pg 26 - Hartsel/Alma
BLUE BARITE

Pg 44 - Marble
MARBLE

Pg 42 - Grand Junction
CLEAR BARITE

Pg 43 - Fruita/Opal Hill
AGATE, PETRIFIED & OPALIZED WOOD

Pg 45 - Ohio City
GOLD, SILVER, *PEGMATITE*

Pg 29 - Hancock/Romley
SILVER, PYRITE, QUARTZ

Pg 43 - Glade Park
CRYPTOCRYSTALLINE QUARTZ, AGATE

Pg 32 - Mt. Antero
FLUORITE, QUARTZ, TOPAZ, PHENACITE, AQUAMARINE

Pg 40 - Ouray Area
CALCITE, DOLOMITE, GALENA, RHODONITE, SELENITE, SILVER, SPHALERITE

Pg 41 - Silverton/Telluride
RHODONITE, (*MINING MUSEUM*)

Pg 39 - Creede/Lake City
WIRE SILVER, RHODONITE, AMETHYST

Pg 38 - Creede/Del Norte
AGATE, CHALCEDONY, PETRIFIED WOOD

Pg 37 - Bona
GALENA, PYRI

Pg 34 - C
EPIDOTE, URA

Pg 3
**GARN
OBS**

Pg 36 - Te
BERYL

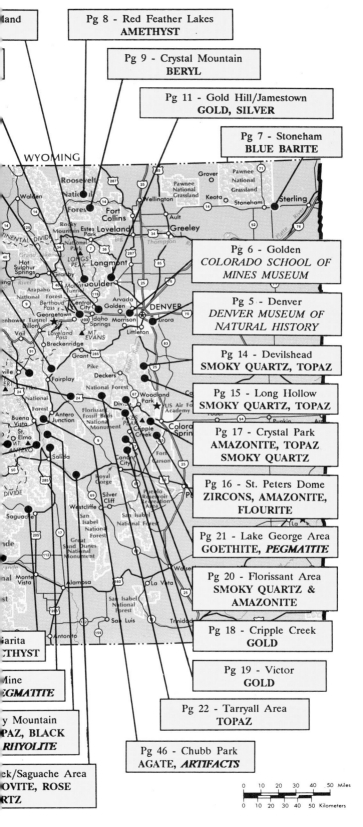

Pg 8 - Red Feather Lakes
AMETHYST

Pg 9 - Crystal Mountain
BERYL

Pg 11 - Gold Hill/Jamestown
GOLD, SILVER

Pg 7 - Stoneham
BLUE BARITE

Pg 6 - Golden
*COLORADO SCHOOL OF
MINES MUSEUM*

Pg 5 - Denver
*DENVER MUSEUM OF
NATURAL HISTORY*

Pg 14 - Devilshead
SMOKY QUARTZ, TOPAZ

Pg 15 - Long Hollow
SMOKY QUARTZ, TOPAZ

Pg 17 - Crystal Park
**AMAZONITE, TOPAZ
SMOKY QUARTZ**

Pg 16 - St. Peters Dome
**ZIRCONS, AMAZONITE,
FLOURITE**

Pg 21 - Lake George Area
GOETHITE, *PEGMATITE*

Pg 20 - Florissant Area
**SMOKY QUARTZ &
AMAZONITE**

Pg 18 - Cripple Creek
GOLD

Pg 19 - Victor
GOLD

Pg 22 - Tarryall Area
TOPAZ

Pg 46 - Chubb Park
AGATE, *ARTIFACTS*

BARITE Blue rose 4x7.5   Hartsel (Orthorhombic 3 H.)

HARTSEL/ALMA

West from Hartsel, beyond the Colo. 9 intersection about one-half mile on the left, take the road heading SW over the small ridge. Proceed about two miles from U.S. 24 to the blue barite claims.

This blue barite is the same mineral discussed on page 8, but in a different habit (shape). The form in which the crystal is usually found is its habit.* All barite is in the Orthorhombic crystal system meaning it will have three axes, all at right angles to each other, but all of different lengths. Barite is very brittle and cleaves (separates) easily. The Stoneham barite is tabular and the Hartsel is bladed, the blades sometimes forming rose-like petals in a circular pattern. Hartsel offers some remarkably fine crystals of intriguing shapes in this pale blue color.

The Hartsel claims are open pit claims, but a small area has been tunneled. Tunneling is dangerous and not recommended. Collectors should dig in the limestone layers for nice clusters, some rose shaped. Single crystals up to two inches wide have been found. On the eastern end of the pit are clusters which look like Christmas trees. Remember to back-fill all holes, and to carry water, as this is a dry, exposed area.

*See *Textbook of Mineralogy*, by E.D. Dana and W.E. Ford, for more on crystal systems.

RHODOCHROSITE (Rhombohedral 3.5-4H.) (right) - Huebnerite penetration (left)

ALMA (Sweethome Mine)

Go SW from Denver on U.S. 285 to Fairplay, then west to Alma on Colo. 9 (6 miles), left at the Texaco station. Up Buckskin Gulch approximately four miles is the Sweethome Mine (the road passes through the mine dump). Collectors have been successful on both sides of the road, but most finds have been on the creek side.

Pyrite and fluorite--both pale green or lavender--and world-famous mineral specimens of rhodochrosite, a manganese carbonate, have been taken from the Sweethome Mine near Alma. The rhodochrosite varies from a pale pink color to an intense pink-red and is found throughout the world in small quantities. There are numerous habits, but one of the most desirable, coveted by museums world-wide, is the rhombohedron from the Sweethome Mine. These specimens are found in single crystals from 1/8" to 2" on a side, in a beautiful transparent to translucent electric red with a pinkish cast. Although this mineral is soft and cleaves readily in three directions, making it difficult to facet, beautiful stones over a carat in weight have been cut from Sweethome material. These stones are not often used for jewelry, but are prized by collectors for their exquisite red-pink color.

The drive from Alma to Kite Lake is a floral treasure house, with outstanding rosecrown and blue columbine. From the Sweethome Mine, continue up Buckskin Gulch Rd. about 1-1/2 miles to Kite Lake. Walk to the mine dump at the base of Mt. Democrat and begin digging on the right side of the dump. Water will help the collector find show-quality cubic pyrite, up to one inch.

DOLOMITE & SELENITE on QUARTZ 6x8   Leadville

LEADVILLE

Leadville, the two-mile-high city, sits among the 14,000 foot peaks in the middle of Colorado. The first attention paid to Leadville was after a placer strike in 1860, but the town lay relatively dormant until 1877. By 1880, there were 15,000 residents, and Leadville was known as the silver capital of Colorado, with H.A.W. Tabor as king. A few of Leadville's more famous mines have been the Matchless, the Black Cloud and the Sherman Tunnel. Today Leadville is home to the National Mining Hall of fame.

Many minerals have been found in the mines around the Leadville area, and while not likely, it is certainly possible to find a specimen on the dumps. Most show-quality specimens have been found underground. Near Turquoise Lake, there is a fee area for hunting turquoise.

The major minerals from the area are barite, dolomite, quartz, gold, pyrite, rhodochrosite, silver and turquoise, all discussed elsewhere. Marcasite, the same composition as pyrite, differs in its internal structure and external form. Cockscomb marcasite forms crystals in groups resembling the red on a rooster's head. Orthoclase, another feldspar (p. 20) like microcline (amazonite), is at this locale usually white, pink, or

MARY MURPHY MINE - Tram Transfer Station

cream with stubby opaque crystals. Twin crystals* are those that have intergrown in a characteristic way, and lucky collectors find twinned orthoclase here. Sphalerite (Isometric 3.5-4 H.), the chief ore of zinc, is usually associated with galena. It varies in hue with the amount of iron, from pale orange to deep red and black. It is often mined for its impurities and is nearly the only source for cadmium.

HANCOCK/ROMLEY

Take Colo. 162 to St. Elmo. The graveyard has some exquisite stone and iron work. East of St. Elmo 1/4 mile, an auto road doubles back to the right leading to the old town of Hancock. For a spectacular 4 W.D. trip, go up to the Alpine Tunnel, about two miles.

Mining at the old town of Hancock was mainly for silver, but pyrite and quartz have also been found. These specimens are small and require a magnification loop for viewing. A 4 W.D. road leads to Romley and up to the Mary Murphy mine, where the aerial tram towers are still in place. Ore was transported from the mine to the mill on this 1883 tramway via 96 ore buckets, each carrying about 200 pounds. The Mary Murphy mine produced over $50-million in gold and silver. Ore samples, pyrite and quartz can still be found on the dumps. The loding chutes are a good place to look for small silver specimens.

*For more on feldspar and twinning crystallography, see, *Colorado Amazonstone* by H.H. Odiorne

RUBY MOUNTAIN

Go south from Buena Vista on U.S. 285 approximately four miles to Fisherman's Bridge campground on the river. Turn left, crossing the river, and take second road on right across railroad tracks. Proceed along base of Sugarloaf Mountain on left. Continue to second mountain which is Ruby Mountain and park on west side.

The Nathrop area, in addition to its attractiveness as a gem collecting site, offers a spectacular view of the Collegiate Peaks. Fishing along the Arkansas River is some of the best in Colorado, and there are raft trips available through local companies. Approaching Sugarloaf Mountain, the gem hunter first sees spots which have been pounded out in the gray/pink rhyolite, about arm height. Be sure to wear a hardhat when climbing to the base of the cliffs, as rock falls from the top. The rubble of rhyolite all along the west side of Ruby Mountain is not the mineral specimen to collect; only the matrix of that mineral. Garnet and topaz positioned in one-inch diameter druse quartz vugs are the main attraction. Rhyolite is a gray, extrusive rock, similar in composition to granite. The rhyolite here is very porous, requiring a point and sledge hammer to break it open and find the tiny vugs.

The rhyolite from this area shows streaks or bands, indicating that it moved while still molten. (Pumice has the same texture and composition.) Sometimes the flow cooled so rapidly, crystals did not have a chance to form; obsidian with a glassy texture was the resulting rock. Black obsidian, "Apache Tears," is found on the north end of Ruby Mountain. Apache Tears "tumble" or cut into attractive gems and are easy to spot in the gray-white dust. The black translucent to opaque stones are usually about pea size.

Spessartine garnet (Isometric 7 H.), dark red to nearly black, and topaz are found here. Most are under 1/4" and are found in vugs 1/2" to 2" in the rhyolite. The garnets almost appear to have been faceted, so sharp are the crystal faces. In fact, there is little change in brilliance after faceting. The garnets are usually only half formed on the matrix, but are very fine specimens. There are six species of garnet and all are valued as mineral specimens and gem stones. *Almandine* and *spessartine* are the most common, varying in color from orange to red. *Grossular* are various colors from pink

GARNET .8   and TOPAZ 1.3   in RHYOLITE - Ruby Mt.

to green. The green variety, tsavorite, can look just like an emerald. *Demantoid* is another green variety and the rarest of the garnets. *Melanite* is the black variety of the species *andradite* and looks like jet but is much heavier. *Uvarovite* is usually small green crystals. *Pyrope*, a fiery red, is seen most often in antique Bohemian jewelry. The *rhodolite* variety is two parts *pyrope* and one part *almandine*, a lovely violet color. The *garnet* group can be found in any color except blue, but the dark red *spessartine* from Ruby Mountain is most often considered garnet. Garnets are common throughout the world.

This topaz, if it has not been exposed to the sun, will be a sherry color; if it has been exposed, it will be clear. Ruby Mountain topaz is small (1/2 inch) but very sharp. Faceted stones larger than one carat, left sitting out for more than seven years have retained their color when not stored in direct sunlight.

An old Indian quarry chipping ground of a red and yellow agate outcropping lies east of Ruby Mountain about two miles. Turn right off Chaffe Co. Rd. 187 just before the white farmhouse, and continue two miles. Close fences, since this is a grazing area.

FLUORITE w/ aqua penetration; SMOKY QUARTZ w/ phenacite crystal top
MICA w/ aqua penetration; AQUAMARINE long crystal 1x2.5

MT. ANTERO

Drive south of Buena Vista on U.S. 285 to Nathrop. Go west four miles on Colo. 162 past the Mt. Princeton Hot Springs Lodge. Go by Mt. Princeton campground along Chalk Creek for six more miles to New Alpine sign on the right. Look for the old loading dock on the left with a sign pointing to Mt. Antero 7-1/2 miles (a good hour drive by 4 W.D.). After a lumpy four miles up through the trees, head left across Baldwin Creek for the top of the mountain. Another mile is timberline and the start of the switchbacks. At the 12,500-foot level is "the island" with parking. Slip around the upper part to the left (north) and the snowfield will come into view. The edges are a favorite hunting area for many collectors, but crystals could be anywhere on this upper rocky expanse. (This area can also be reached by going down from the parking area at the 14,000-foot level.) Continue up the last long switchback to the saddle, about 13,000 feet. After 1/4 mile the road forks; right fork goes to Mt. White or back west to the California Mine, about two miles. The left fork continues on the south side of Mt. Antero 'til the road again forks. Take the left road the last few switchbacks to the parking area at the 14,000-foot level.

The view from Mt. Antero, probably *the* classic collecting area in Colorado, is like the view from the top of the world. South are White Mountain and Shavano, another 14,000 foot peak. North are Princeton, Yale and Harvard, all 14,000 footers. To the east is the valley with Chalk Creek running down to the Arkansas River, Ruby Mountain and the town of Salida. West is Baldwin Lake and the road to the top.

FIELD COLLECTING ON MT. ANTERO

The aquamarine* in this gem hunter's paradise is usually light blue, with crystals from needle size on up. The first published report about the area was in 1887 by George F. Kunz. Collectors have been coming up Little Browns Creek to comb the area since and are still finding crystals.

To begin the hunt, find some trace of blue color and follow it, looking for the source. The most common finds are aquamarine; both clear and smoky quartz in beautiful, small plates as well as large single crystals; topaz (clear and sherry); phenacite (clear to yellow); fluorite (green and purple); and a number of less common minerals. It would be possible to have one or all of these on a microcline plate, complete with tiny mica books. Finds are infrequent at this locality but usually worth every effort made for these excellent specimens. This is the classic locality for our state gemstone, aquamarine. It is about an hour's walk from this site to the top of 14,269-foot Mt. Antero.

Travelers choosing the right fork (just above the saddle about 13,000 feet) can go to the California Mine dump or to Mt. White. Aquamarine, clear beryl or light green beryl, molybdenum and smoky quartz are on the dump of the California Mine. These mineral specimens are beautiful and often found with multi-minerals on the same specimen. Small chunks of clear cutting material, usually beryl and smoky quartz, may be found at any of these locations. At Mt. White, park where there are some road cuts in the mountain seen from the saddle. Search on the road cut side, and also on the back side of the mountain for aquamarine, smoky quartz and topaz (usually associated with quartz).

*for other colors of beryl see p. 9

EPIDOTE, longest 1.5x2   Calumet Mine (Monoclinic 6-7 H.)

CALUMET MINE

Just before crossing the river into Salida from the northwest, turn left and take the Ute Trail Road (Chaffee Co. 175) approximately seven miles to Chaffee Co. 185. Continue about another mile to the quarry on your left (north). Take the right fork, (not toward Turret) and continue about 5/8 mile to a small stand of aspens on the right and the remains of the old loading dock. Up the hill is the open pit of the Calumet Mine. Park and follow the trail on the north side of the gully past the old powder house to the pit area.

Single epidote crystals from the Calumet Mine area measure up to two inches on a side. Epidote is often associated with white quartz, and although not facet quality, does make excellent mineral specimens. Small iron garnets and magnetite crystals can also be found here. **Do not** enter the mine shaft; there is a 25-foot drop at the end of the old wooden steps going down into the mine.

Along Chaffee Co. Rd. 184 from the quarry, a little over two miles, is the Turret area which has some private cabins and mining claims. This is a pegmatite district containing beryl, columbite, corundum, garnet, graphite, hematite, magnatite, microcline, muscovite and quartz. Any of the above minerals found in this area would probably be very small and used only for a reference collection. Remember--permission is required before collecting on any private claims.

URALITE 6x7.5   Calumet Mine

Uralite embedded in calcite is found on the dump west of the old mine entrance. The specimens are large chunks of white calcite with green penetrations. The uralite must be leached out of the calcite with hydrochloric acid. Uralite is an amphibole pseudomorph after pyrozene, and this area produces some of the best specimens in the world--up to two-inch crystals. A pseudomorph is a "false form," created when a mineral is altered but the external form is preserved. An example is petrified wood, where the original fibers or cells have been replaced by silica. The squarish gray-green uralite looks extremely attractive set against the white of calcite.

On the top of the pit is a white limestone layer containing the black-green epidote. Two-hundred feet uphill are also old diggings. The dump area in the pit has produced some multi-mineral small plates, mainly epidote crystals, which could have tiny garnets, magnetite or quartz with the epidote. This is a favorite collecting area for students at the Colorado School of Mines. Across the valley is a spectacular view of some 14,000-foot peaks, including Mt. Antero, which often has interesting cloud cover. Be sure to take water to this area, and keep a watchful eye out for snakes, as rattlers have occasionally been spotted here.

ROSE QUARTZ 6x8   - Devils Hole

TEXAS CREEK/SAGUACHE AREA

Travel west on U.S. 50 from Canon City 27 miles to Texas Creek, or go east from Salida. Cross the river and railroad tracks to the north and follow the dirt road up the draw. Do not take the first right, but continue up the hill. The road swings back to the right over the hill and down to some cabins. Beyond the cabins, rose quartz covers the southern slope of the hill to the left. The road just left of the gully leads to the mine, a total distance of six miles.

DEVILS HOLE - This area is well known for beryl, columbite-tantalite, microcline, muscovite, and rose quartz. Imbedded in the basic rose quartz pegmatite are beryl crystals 12 inches in diameter and up to 24 inches long. The pale green to blue-green opaque crystals will not be clear, but there are areas clear enough to get some faceting material. On the right side of the pit is the best faceting quality rose quartz, which is the main mineral in the area.

Also in this area are some very large muscovite mica books. Micas have perfect cleavage and flexibility of individual cleavage sheets. Crystals are monoclinic but often look hexagonal. The various mica minerals (muscovite, lepidolite, biotite, etc.) are so similar in internal structure that several species may crystallize in parallel position. Muscovite is pale or colorless, and large sheets were used in the early mining days for window panes. They cleave readily and are fun for children to "peel." The black columbite-tantalite crystals can easily be spotted in the quartz, usually close to the large beryl crystals.

MINE BUILDINGS - Bonanza Area

Travel south on U.S. 285 from Poncha Springs to Villa Grove. Turn right up Kerber Creek just before entering Villa Grove and travel 13 miles to the Bonanza Mining District. Check before entering private property in this or any area.

BONANZA - Galena, quartz, pyrite and rhodochrosite* exist on the old mine dumps in this region. The specimens from this area are generally reference collection quality and usually rather small. Old 4 W.D. roads take the traveler by many old mine buildings. There is an interesting graveyard just out of town.

SAGUACHE--South from Villa Grove is the mountain range containing seven of Colorado's 14,000-foot peaks.** At the southeast is the Great Sand Dunes National Monument. Go west on Colo. 114 just north of Saguache to the Thunder Egg area (see map p. 38). Some of these geodes will be lined with amethyst crystals.

LA GARITA - Near La Garita (Biedell Gulch area), phoenix terminated amethyst quartz crystals, some with two-phase water bubble inclusions, have been found. The area has been destroyed by gold mining, but excellent specimens can be found in old collections.

*See index for previous discussion of these minerals.
See **Colorado Traveler, *Mountains and Passes*

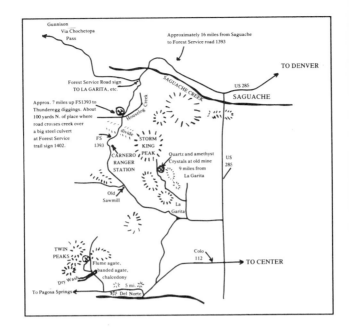

CREEDE/LAKE CITY

Go south from Saguache to Del Norte, then northwest to Old
Women Creek area.

DEL NORTE--Along the draws, creeks and hillside in
this area are golden plume agate, dendritic and banded
agate, blue chalcedony and petrified wood. The silica
group can crystallize in many distinct forms, the most
common being quartz. It occurs in all colors and a huge
variety of habits. Quartz is often included with other
interesting minerals, such as tourmaline incased by a
clear quartz. Aventurine is a green quartz with tiny
flecks of mica that give it a sparkling effect.

When quartz is very fine grained, it is called
cryptocrystalline. The following familiar names are
cryptocrystallines: agate (moss-like inclusions or banded); bloodstone (green chalcedony with red spots);
carnelian (red-orange); chalcedony (fibrous and sometimes banded); chert and flint (dark, massive, usually
nodules); jasper (dull red or brown); onyx (black);
sardonyx (banded in black or brown and white). The
terminology in just this *one* mineral category is some-
what a challenge - agate is always quartz but quartz is
not always agate! While all minerals have specific and
distinct terminology, quartz is most common and so
perhaps has the greatest vocabulary.

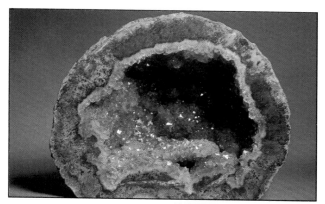

AMETHYST GEODE 10 diameter

Travel west from Del Norte on U.S. 160, 31 miles to South Fork. Take Colo. 149 northwest 22 miles to Creede.

CREEDE is primarily a silver mining town today. In early days, it was the home of such famous mines as the Bull Dog, Amethyst Queen and Commodore. Wire silver comes from this area and small specimens of cerussite, chlorite, limonite, malachite and even wire gold can be seen with a magnification loop. Sow belly agate from the Amethyst Queen Mine is unique to the area. Sow belly is bands of wavy white, blue and lavender chalcedony (quartz) interspersed with amethyst crystal bands. When sliced, it looks like bacon, but with nicer colors. The Amethyst Queen is also the site of nice facet quality small crystal plates of amethyst.

Leave Creede southwest toward Spar City and begin climbing the plateau. Travel almost 50 miles, over Slumgullion Pass, and into Lake City.

LAKE CITY--Thousands of dollars in gold and silver have been mined in this rich area. Gold, copper, lead and rhodochrosite in quartz vugs all come from here. Any mine dump could be a treasure house. There are several old mining towns in this region. In 4 W.D., the route by Lake San Cristobal and over Engineer Pass, down to Animas Forks offers beautiful rhodonite (5.5-6 H.). This is not crystalline but produces chunks big enough for bookends. These chunks could be made into cabochons: rounded convex tops with a flat back, in round or oval shapes, or (rarely) in freeforms. Cabochons were the first formal cutting of gem material before the birth of Christ. They are especially suited to agate, opal and jade or any of the opaque stones which will produce a star like a star ruby.

WHITE QUARTZ group 5x6.5   Camp Bird Mine (Rhombohedral 7 H.)

OURAY AREA

Travel south from Ouray on U.S. 550 to Colo. 361. About five miles are the dumps of the Camp Bird Mine on the left. Leave the Camp Bird and return to U.S. 550 south to Silverton. There are many mine dumps on the east side of the road with small but good specimens. This area can be reached without 4 W.D.

OURAY--Minerals abound in the mines and on the dumps in this famous Colorado mining area. Here, the careful collector can find calcite, dolomite, galena, gold, manganocalcite, quartz, pyrite, rhodochrosite, rhodonite, selenite, silver and sphalerite. Most collecting is done on the mine dumps, subject to permission of the mine owner.

Calcite, like quartz, is found all over the world in many forms and colors and in combination with many other minerals. Its most common form is limestone and marble which make interesting stalactites and stalagmites. Crystals are often twinned and show more diversity in form than any other mineral (see also p. 42). Camp Bird Mine manganocalcites (manganese mixed with calcite to create a dreamy pink color) are sought by museums world-wide. The rhodochrosite, also a carbonate mineral containing manganese to give it the pink color, is usually opaque and very different from the ones discussed on page 27. This rhodochrosite is baby pink rather than the intense pink of the Sweethome Mine specimens, but both are prized. Dolomite, another carbonate often found with calcite, is usually cream to gray-white, and like rhodochrosite, 3.5-4 H. All three of the above crystals are Rhombohedral in shape.

MANGANOCALCITE on GALENA (Isometric 2.5 H.) 5x10   Camp Bird Mine

The metallic-looking crystals of galena, gold, pyrite, silver and sphalerite found in this area are all in the Isometric crystal system. Galena is very easy to recognize with its gray color and cubic shape. It is also heavy - the principal ore of lead. Galena crystals weighing hundreds of pounds are not uncommon in some parts of the world - a nice specimen from this area might measure 1/2" on a side. Sphalerite is most often associated with galena and is the chief ore of zinc. The color varies with the iron content from pale orange-yellow to deep red and black. Cadmium, indium, gallium and thallium metals occur as impurities in sphalerite and it is virtually our only source for them.

From Silverton, head north on highway 110 to the Cement Creek turnoff, which goes to the Gladstone and American Tunnel area. After a stop here, continue north about eight miles to the site of the old Sunnyside Mill at Eureka. From the old mill dumps, go up the road to Animas Forks.

SILVERTON--Animas Forks is a good place to dig for rhodonite.* The pieces have a nice pink color. Because of its 11,500-foot altitude, this area was not worked until 1875. A nice 4 W.D. trip from Silverton back to Ouray is via the Ophir Road over the pass to the nearly deserted town of Ophir.

Travel north on highway 145 to the Alta turnoff and proceed up the road. After investigating this area, continue on to Telluride.

TELLURIDE has a fine mining museum and there are some great 4 W.D. trips from this area: above Bridal Veil Falls to Blue Lake; up to the Tomboy Mine; over the pass and back down to the Camp Bird area. For the stout-hearted, "Black Bear" is a scenic trip. There is a possible "killer" specimen in any of the many mine dumps in this mining district.

*See p. 39

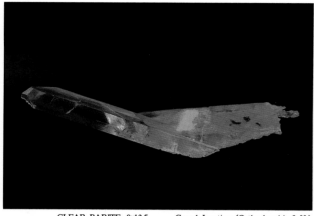

CLEAR BARITE .8x12.5   Grand Junction (Orthorhombic 3 H.)

GRAND JUNCTION

Take road 25 or 29 north of I-70 in Grand Junction past the airport toward the Book Cliffs.

Near the foothills in the Book Cliffs area are brown nodules sticking out of grayish limestone, which range from one to five feet in diameter. Many of them have been broken and the barite removed. To open a whole nodule, a use a point and hammer and open from the side or bottom to avoid damaging the inside. Usually there is a layer of white or clear calcite under the brown rock, with the barite in the center. Note the natural square cleavage of the calcite* (Rhombohedral, 3 H.).

This barite is usually water clear and the area has produced great mineral specimens and facet grade material. Barite is a very soft mineral and cleaves easily with temperature change, so be careful of water temperatures when cleaning. A nice specimen from this area would be thumbnail (one inch) and larger.

Colorado National Monument is four miles west of Grand Junction on U.S. 6 & 50.

The geology of the Colorado National Monument covers a wide span of time. Rocks from the Precambrian era, more than a billion years old, and lava flows from less than a million years ago can be seen in a number of places. The mountain building uplifts cracked the crust of the earth, creating a 10-mile fault. Rainfall here is now about 11 inches annually; but

*See p. 40

PETRIFIED WOOD products

during the Ice Age it was probably much heavier. The Colorado River moved with much greater force then, carving the open caves and monoliths from this crystalline rock area composed of granite, gneiss, schist and pegmatites.

South of Grand Junction is Glade Park, and 3-1/2 miles south is a good collecting locality for various agates and cryptocrystalline quartz.* These materials are usually found in pebble form in a variety of colors.

Travel south from Grand Junction to Whitewater. Turn right on highway 141 for 100 miles to Naturita. Along the washes and creeks in this 100-mile stretch, cabbing material can be found.

Agate and petrified wood are found on Pinon Mesa south of the Colorado River all the way into Utah. There is opalized wood at Opal Hill, two miles south of Fruita and 1/4 mile west of the road to the Colorado National Monument. Common opal is hydrated silica, less hard than cryptocrystalline quartz (5-6 H.), and is anhedral (a crystal having no plane faces at all). It is opaque to translucent, sometimes waxy looking and is often confused with chalcedony. In speaking of opal, the usual reference is to "precious opal" with its riot of colors on either a white or black background. Common opal is the same mineral without the aurora borealis of color. All of this material takes a magnificent polish and has beautiful color, though solid and without flashing. Often, opalized wood will still have the bark attached, and the tree rings apparent, making specimens most interesting.

*See p. 38

WHITE MARBLE slabs (notice drill grooves)

MARBLE/OHIO CITY

After spending the night in Glenwood Springs, enjoying the natural hot springs and swimming pool, travel south from Glenwood Springs to Carbondale. Turn on Colo. 133 along the Crystal River. Stop in Redstone to see the Inn and cold coke ovens. On the approach to McClure Pass, turn left and proceed to Marble.

Along the road approaching Marble are large pieces of white marble which probably fell from train flatcars in the mining heyday. Around the old cutting mill and smelter area are large slabs and pieces. From town, it is four miles to the quarry, which opened in 1890 and closed in 1941. Marble for the Lincoln Memorial (worth $1,070,000) was shipped by the end of 1916. The Tomb of the Unknown Soldier material was also mined at this quarry. In evidence today is the section from which the solid block for the Tomb was mined. Moving the slab down the hill to the mill for cutting was apparently no easy chore, for the block weighed 55 tons and took a year to remove. The old Colorado State Museum and Colorado National Bank buildings in Denver are also made from this beautiful material. Marble* is a coarse-grained metamorphic rock of limestone or dolomite. This area's marble is extremely pure and white.

Toward Schofield Pass (4 W.D.) a few miles out of Marble is the old mining town of Crystal. The abandoned mill along the Crystal River is featured on many calendars.

---

*See p. 40

MUSCOVITE 8.5x12   Ohio City

Proceed over McClure Pass, along Paonia Reservoir. Turn left at the Kebler Pass Road and continue over the pass to Ohio City, a nearly extinct ghost town.

Gold and silver were both taken from the famous Quartz Creek (Gold Creek) collecting area. Scores of pegmatite bodies, rich in lepidolite and muscovite mica exist in this area. Muscovite mica (discussed on page 36) is the smoky to clear variety. Lepidolite has lithium in it and is usually pink or lavender. This area has only small examples; large massive* hunks are in the Pala Mountains of California. Micas are often found in the same pegmatites as beryl. Beryl from this area is not well crystallized, nor is it clear.

Topaz and many other minerals were collected in the old days from this mining district. The silver and gold rush struck here in the 80s and 90s. Today, owners charge a small fee for collecting in the best areas, but there are many pegmatites not under claim. Always get permission before collecting on private claims.

From Ohio City, travel to Crested Butte, rich in mining history, and with some contemporary working mines. Past the ski area is the town of Gothic, another ghost town with a view like Switzerland. In Gothic, and towns like it, miners braved high altitude snow in bitter cold, living in uninsulated cabins and working a 12-hour shift, all for $3.00 per working day. Some became very rich but most did not, and much of the great treasure of the Colorado mountains has yet to be found.

*Term used in gemology, with no reference to size.

ARROWHEAD (JASPER), DENDRITIC OPAL - Chubb Park

CHUBB PARK

On U.S. 285 south of Denver and Fairplay is Trout Creek Pass. The next four miles, nearly to the bridge and Chaffee Co. Rd. 303, on the west side of the highway, is an area for agate, arrowheads and opalized wood. Most of the artifacts are located in the washes and open spaces where the wind may have uncovered them. This collecting area extends from the highway down to the creek and up the other side.

There are numerous areas to hunt agates and artifacts in Colorado. Large boulders of petrified wood were found during excavation for a phone company building addition in downtown Denver. Near Parker, east of Denver, nearly any lot and many road right-of-ways will yield specimens. In this book, emphasis has been on locations of crystals, bodies with crystal faces as external evidence of their regular internal structural arrangement. Massive minerals are those with no external indication of an internal crystalline nature. The amorphous minerals are those without crystalline structure (like glass). The cryptocrystalline (p. 38) are minutely crystalline or indistinctly crystalline but not discernable under magnification.

The dendritic cryptocrystalline piece pictured above is a wonderful example of inclusions. Inclusions are minerals which formed along with the main mineral (quartz in this case) or pre-existed and became trapped inside the quartz. In this case, deposits of minerals moved along thin fractures within the quartz and became trapped in *tree-like* shapes; thus the name *dendritic* quartz. There are many beautifully patterned included quartz specimens, some with specific names.

AUTHORS ON MT. ANTERO

Lee and Tag McKinney have spent long weekends and vacations for the last ten years collecting in the Salida, Buena Vista area and their own mining claims. Still, many mineral specimens remain to be unearthed here and throughout the state.

When exploring any field collecting site, the best excavating prospects are existing holes. Inspect the dump (loose material around the holes) for partial crystals. How deep is the rock? What color? Does it have mineralization on it? Observe the surrounding area and previously disturbed spots. Children, with their sharp eyes (and proximity to the ground) are often the best novices.

FIELD TRIP CHECK LIST

rock hammer	field boots/gloves
crack hammer/sledge	hat & hard hat
chisels/jack hammer pt.	safety glasses
prybar/gad bar	rain jacket/sweater
pick & shovel	long pants
3-prong garden scratcher	shirt with pockets
screwdriver/probe	sun screen
dental pick/nut pick/nail	bug spray
brush/whisk broom	water/lunch/snacks
screens/seives	maps/field guide
flats/packing/bags	notebook & pencil
egg cartons/baggies	first aid kit
back pack/fanny pack	camera/binoculars
hand lens/flashlight	personal medication

Index